YOUR KNOWLEDGE HAS VALUE

Marco Alexander Caiza Andresen

Evidence Based Reasoning / Statistical Literacy Teaching Statistics and Econometrics

GRIN Verlag

Bibliografische Information der Deutschen Nationalbibliothek:

Die Deutsche Bibliothek verzeichnet diese Publikation in der Deutschen National-
bibliografie; detaillierte bibliografische Daten sind im Internet über http://dnb.d-
nb.de/ abrufbar.

Imprint:

Copyright © 2006 GRIN Verlag GmbH
Druck und Bindung: Books on Demand GmbH, Norderstedt Germany
ISBN: 978-3-638-77602-8

This book at GRIN:

http://www.grin.com/en/e-book/73336/evidence-based-reasoning-statistical-literacy-
teaching-statistics-and

GRIN - Your knowledge has value

Der GRIN Verlag publiziert seit 1998 wissenschaftliche Arbeiten von Studenten, Hochschullehrern und anderen Akademikern als eBook und gedrucktes Buch. Die Verlagswebsite www.grin.com ist die ideale Plattform zur Veröffentlichung von Hausarbeiten, Abschlussarbeiten, wissenschaftlichen Aufsätzen, Dissertationen und Fachbüchern.

Visit us on the internet:

http://www.grin.com/

http://www.facebook.com/grincom

http://www.twitter.com/grin_com

Universität Paderborn
Fakultät für Wirtschaftswissenschaften
Department 4: Economics
Ökonometrie & Statistik

Statistisch-Ökonometrisches Praktikum
WS 2005/2006

Evidence Based Reasoning / Statistical Literacy
Teaching Statistics and Econometrics

Marco Alexander Caiza Andresen
Course of Studies: Master in International Business Studies

Paderborn, January 12th, 2006

Contents

I.   Introduction

Outdated education models, technological advances and increasing enrollment of students have led to involve Web-based education in some economics classes of various universities. The options range from Web-based applications in traditional classes to complete online courses without any face-to-face contact. Two facts are stressed with special regard to statistics and econometrics classes in this paper. These are, firstly, the problems tutors[1] have to teach students the essential contents of the courses (this refers also to many aspects of economic undergraduate courses) and, secondly, the problems tutors face to find the right way to teach by using the possibilities the technological advance offers to education methods.

1.1   Purpose of the Paper

Given that only a few written sources on teaching statistics and econometrics exist (Becker and Greene 2001) and having in mind that econometrics is part of the economics education, articles regarding research in overall economic education will also be analyzed in this paper. Due to the latest articles the aim of the paper is to compare and combine the findings of different studies carried out in order to find the best way of teaching econometrics and statistics. After this short introduction the main part of the paper gives an overview of the conventional way of teaching statistics and econometrics and indicates the problems involved. This is followed by a section on new initiatives in the education of econometrics and statistics. Thereinafter, three forms of teaching – traditional (off-line) instruction, hybrid Internet-based instruction and Internet-based instruction – will be dealt with to see in how far the proposed initiatives already have been applied on the subjects. In the last part a conclusion is drawn to summarize the main findings and to show the direction of future teaching in this field. As it already became clear in the headline subject matter of this paper is the aspect of teaching and not learning (which will be analyzed by a fellow student). Thus, all aspects of learning statistics and econometrics, such as the Ten Principles of Learning Statistics developed by Garfield (1995) or the study of Johnson (2005), are omitted; the work deals exclusively with the perspectives of the teaching institutions and not of those on the receiving end of the instruction.

---

[1] A survey by Becker and Watts (2001) shows the dominant picture of the US undergraduate economics teacher in higher education institutions (without doctorate universities). This teacher is male, Caucasian, has a Ph.D. degree and lectures to his class by using the chalkboard and a standard textbook. He spends 40 percent of his time on teaching and the same amount of time on research.

II. Main Part

2.1 Main Problems in Teaching Statistics and Econometrics

Sowey (1983, p. 257) defined econometrics as "[…] the discipline in which one studies theoretical and practical aspects of applying statistical methods to economic data for the purpose of testing economic theories (represented by carefully structured models) and of forecasting and controlling the future path of economic variables." Thus, it is not enough to provide the students with the theoretical knowledge, it is also necessary to give them appropriate practical examples so that they can use the theoretical key concepts for quantitative analyses on their own. In opposition to that stands the fact that Principles instructors in economics spent most time in class lecturing, leaving insufficient time for practical activities (Becker and Watts 2001). Becker and Greene (2001) analyzed the essential topics to be taught to undergraduates in statistics and econometrics and additionally point out the problems in the traditional instruction. Next to problems related directly to specific statistical topics, there are also general problems: (i) abstract and dry textbooks and (ii) use of problem sets[2] of made up data and unrealistic numerical examples. Although an immense supply of statistic textbooks exists, there is little attention paid on the applications of concepts and procedures. To engage students more in class it is necessary to use real-world examples, which can be obtained from history, news, popular culture, the classroom itself and the students' lives. Especially current events in the news can be used to show the importance of economics and statistics in real situations (compare also Hansen et al. 2002 and Hamermesh 2002). Articles on active learning techniques, as published in the Journal of Statistics Education and the Journal of Economic Education (JEE), can also help to teach more actively in classroom. The high importance of mathematics in statistics and econometrics in comparison to other economics classes may deter some students from signing up because instructors view students' skills in numerical calculations and algebra as extremely important, in graphs as important and in calculus as to some extent important (Becker and Watts 2001). Becker and Greene go on with a detailed description of the necessary undergraduate concepts and skills to be taught, which seem to be difficult to many students. These are probability, sampling and sampling distributions, hypothesis testing, regression to the mean, motivating the least squares estimator and alternatives to least squares. Nonetheless, basic subject in any statistics and econometrics undergraduate course has to be calculation and use of descriptive statistics

---

[2] Although teachers mostly develop their own problem sets these are rarely based on press readings or on scholary publications (Becker and Watts 2001).

(mean, median, standard deviation, etc.), as well as teaching of basic skills related to data management, computation and graphing. The more difficult concepts will be introduced here in shortform separately. Concerning probability many students are able to repeat basic formulas and rules but the distinction between marginal, joint and conditional probabilities in applications are from time to time difficult, even for instructors. Here the authors recommend sources of examples for such cases, such as Marilyn vos Savant's weekly Parade magazine column and a book by Paulos (1995). Regarding sampling students often have problems to understand that from sample data calculated statistics used to estimate corresponding population parameters are themselves random, with in the sampling distribution of the statistic, which is a histogram, represented values. Due to the importance of this concept, it is inappropriate to let the students work with an imaginable construct while they could develop a histogram of possible values of a sample statistic themselves through group experiments in computer labs. One important thing they would learn by experimenting on their own, is the difference between the law of large numbers[3] and the central limit theorem[4]. Thus, students could see how a standard normal random variable (with mean of zero and standard deviation of one) that does not degenerate to a single value, as the sample size increases infinitely, is created through the standardization of a sample mean. For the students in the computer lab bootstrapping is a natural real-world extension of their work with sample distributions. Here from the original sample repeated samples are taken (with replacement) and from these the distribution of the desired descriptive statistic is deduced. Without requiring further assumptions about the underlying distribution of the population or the context of the considered real-world problem this sampling distribution is used to construct an interval estimate of the population parameters of interest. Consequently, the use of bootstrap as a teaching tool provides students with early and practical experience to a strong research tool. Regarding hypothesis testing many students have problems with the understanding of the tradeoffs between Type I[5] and Type II[6] errors. Also the debate over the application of statistical significance in opposition to magnitude and practical importance of an effect is ignored. Most econometrics textbooks recommend stress on statistical significance and minimum concentration on the size of the estimated effect. Confidence intervals could be used to stress the importance of sign, size

---

[3] The law of large numbers means that by increasing the size of the sample, the sample mean coverges to the true mean.
[4] The central limit theorem states that for many samples of like and sufficiently large size, the histogram of these sample means appears to be a normal distribution.
[5] A Type I error arises when the the null hypothesis is incorrectly rejected.
[6] A Type II error arises when the null hypothesis is not rejected when it is in fact false.

and significance, thus, avoiding the use of a formal null hypothesis. Especially the conditional nature of a Type II error is difficult to understand: hence, here again, the practical use of computers could show the students the changing size of the Type II error by increasing the sample size. On the subject of regression to the mean a fallacy is that relatively high values are expected to fall toward the average while relatively low values are expected to rise to the average. This often occurs when analyzing high-value and low-value points of a splitted sample separately. If regression to the mean in fact exists, the variance of the distribution as a whole should decline and all values should cluster continually closer to the mean. Looking at averages of subgroups does not give an idea about whether the variance of the entire distribution has declined. About motivating the least squares estimator scatterplots usually have provided the means for introducing linear least squares regression. Nonetheless, it does not remain unquestioned that the suggestion of minimizing the sum of the squares of the residuals should be the correct instrument to estimate a mean value of a dependent variable conditional on the values of the independent variables. Although there are some approaches to motivating the least squares estimator, many students still do not see an obvious reason why it should estimate the relevant population parameter. Thus, the authors suggest estimation on the method-of-moments, which is the norm in advanced statistical treatments. This method is based on the understanding that a population parameter estimator comes from the corresponding feature of the sample, in other words, estimators of population parameter must have sample properties that mimic similar properties of the population model. Thus, within the method-of-moments method the starting point is not the data but the properties from the population. The approach states that from the population model generated sample data must have the same properties, as forced on the data due to selection of the intercept and slope of the sample regression line. At last, traditional econometrics course primarily emphasize the algebra of least squares estimation of parameters in well structured models. Due to advances of incorporating computer labs in econometrics education nonlinear modeling and nonparametric regression techniques should be applied because it is important to show the students alternatives, although they must be able to handle least squares. Students now can use these tools without learning difficult programming syntax or mathematical specialties.

Sometimes ideas at the frontier of research (e.g. topics the teacher is temporarily working on) presented in class can reveal the excitement of the subject and can attract the students who consider become majors (Hamermesh 2002) but it should be considered that most of the students will not take any other courses than Elemtary and Principles in these subjects and that many of them have very limited analytical abilities (compare Case 2002). Altogether, not too many and too advanced contents should be taught and these carefully and real-life oriented. Finally, Hamermesh (2002) introduced some general hints on teaching all kinds of economics courses of which the most important ones will be presented shortly. Besides the lecture notes the instructor should have detailed notes with illustrations and examples worked out that should all be ready several classes ahead of schedule to avoid last minute preparations. It also should be noticed that with Power Point, although the complete lecture is ready-made for presentation, passive learning is encouraged by giving the students an incentive to avoid active engagement with the material as they will certainly rely on the printed notes of the lecture. Despite the vast amount of content included in each lecture, there are typically only three or four main points, each briefly expressible. These should be listed for the students at the end of the lecture to provide them with a good overview. A detailed syllabus can be seen as a contract between instructor and students and eliminates possible misunderstandings. Finally, two midterms and a final exam should be enough and students should be offered a question-and-answer session outside the normal class time before each exam. There are also some other important and interesting hints for teaching economics but as these are more general it is recommended (especially for teachers) to directly read the article of Hamermesh (2002).

2.2   New Initiatives in Teaching Statistics and Econometrics

The previous part demonstrated the necessary contents for econometrics classes while in the following sections the focus will be set on how to teach econometrics and statistics classes independent of the subject's theoretical contents. To sum up once again, the lack of real world topics for applications in the textbooks as well as in classes is a problem that has to be dealt with intensively. The theoretical concepts shown above and their application must be focused by the instructors. Although nowadays it is possible to integrate the use of computer labs in education, in teaching business, econometrics and economic statistics the "chalk-and-talk" mode still dominates; thus, modes should be developed to incorporate computer technology in the most efficient way into classes. Nonetheless, changes can only be expected if students, instructors and education institutes are open to it (Becker and Greene 2001). This starting points go along with the initiatives for improving economic education in general. Salemi et al. (2001) introduced five new initiatives for research in economic education initiated by the Committee on Economic Education (CEE) of the American Economic Association (compare also the improvement suggestions by Hansen et al. 2002). These are Teaching Methods and Incentives for College Level Economics Instruction, Ph.D. Education in Economics in the United States, Improving the Assessment of Student Learning in College Economics Courses, Long-Term Effects of Learning Economics and Efficiency in the Use of Technology in Economic Education. Each initiative should be executed by a project group of scientists. The initiatives could be divided up into three categories. The first category deals with the direct in-class improvement through more efficiency. For that the task of the project of assessment is the development of test instruments to examine if students really learn more effective when tutors use an active learning method as a substitute for standard lecturing. The technology project should show the effects of use of information technologies in class because an evaluation of the costs and benefits of electronic technologies is important since there is still little evidence on the learning- and cost-effectiveness of recent changes in technology and the lasting-effects project should recommend strategies for ensuring that the students are able to use their knowledge learnt in the economics classes even years after course completion. The second category deals with different ways to encourage the changes in teaching. The teaching-methods project should establish a certificate of achievement in economic education, as well as workshops to initiate a more active learning concept. Furthermore, a new section of the Journal of Economic Education should be initialized through which the assessment project tries to find out if tutors will improve the quality of their tests and assessment practices when they get a publication opportunity. In the last category the graduate-training project should help to determine the reasons and effects of the upcoming decrease in economics Ph.D. recipients. Here, in this paper, the fist category is being focused. Thus, the next part shows the different offerings in statistics and econometrics teaching models.

2.3   Models of Teaching Statistics and Econometrics

Unfortunately, two of the studies used in this paper compared only two of the models at a time and only one study compared all three. These studies are the studies of Ward (2004) comparing a hybrid and a traditional model of Elementary Statistics and of Utts et al. (2003) comparing a traditional and a hybrid class of the Introductory Statistic course concerning the student's performances and attitudes. Brown and Liedholm (2002) compared a Web course, a hybrid course and a traditional course in Principles of Microeconomics concerning the performances of the students. Eventhough this is a microeconomics course and due to the lack of further data concerning the different teaching approaches of specific statistics and econometrics courses, especially of a completely online class, it is used in this paper in accordance with the responsible teaching professor at the university. In the following the characteristics of the individual courses will be introduced.

2.3.1 Traditional Instruction

The so-called traditional instruction sees professors and students in the same location at the same time whereas instruction is face-to-face (Ward 2004). Generally, computer-based components are not integrated. In some ways the strict traditional "chalk-and-talk" is seen as outdated as many scientific courses at universities renew their teaching models, involve new components, use new media, etc. The course designs of the traditional classes of the three studies are analyzed as followed. The traditional course in Utts et al. (2003) includes ten weeks of class with a 50-minutes lecture held three times a week (with 200-250 students) and a discussion section of 50-minutes (40 to 60 students) once a week . The students had to hand in weekly homework, write three midterm and a comprehensive final exam. Basis of the class was a textbook and there was no use of online or interactive materials because although such a type of course would provide the benefits of online interaction and immediate feedback, while still allowing sufficient class time for discussion and explanation, the instructors thought that the students may be overwhelmed by the available resources and would not use any of them to full advantage. The traditional model of Ward's (2004) course also met three times a week for fifty-minute classroom sessions. Four class sessions took place in a computer lab where students used a statistical computer package to analyze data they had generated in class or worked on their final project.

The ordinary classes included lecturing, answering questions, interactive worksheets, collaborative problem sessions, calculator activities, tests and quizzes. The material used contained lab exercises, worksheets and problem sets identical to the documents on the course Website. Power Point reviews, applet demonstrations and recommended readings, as used in the hybrid class, were also made available to the traditional class. Communication between instructor and students was possible by telephone, email or during office hours. The traditional course model taught by Liedholm (Brown and Liedholm 2002) met face-to-face three class hours a week. The instructor engaged the students through the use of animated Power Point slides, videos, group demonstrations and calling on individual students. The basis for the class was also a textbook, which was used in all three modes of the course, as well as multiple-choice in exams and e-mail and course Web sites for communication was also the same.

2.3.2 Hybrid Internet-Based Instruction

Hybrid instruction utilizes both distance learning via the Web and the traditional classroom face-to-face format in combination (Utts et al. 2003). However, many different types of hybrid courses exist, as shown in Ward (2004). These can be held in a computer lab with students completing online activities in the presence of their instructor (see Levine 2002), with face-to-face contact only at an orientation meeting and the final exam (see MacGregor 2001), with several face-to-face meetings with the instructor to discuss problems that were not solved by electronic contact (see Yablon and Katz 2001) or by offering the students a selection from various online and in-class activities (see Rensselaer Polytechnic Institute 2001). The designs in the three studies are the following. The hybrid Web-based class of the Utts et al. (2003) study met once a week for 80 minutes of which about a third of the time was used for a quiz regarding last week's material and the rest of the time to provide the students with an overview of the material for the following week and to demonstrate some material on the Web. The original idea about this class was that the instructor should motivate the students and explain the concepts, so that the students could work out the details using textbook and online materials on their own. There were also a midterm and a comprehensive final exam, as well as weekly homework. Utts taught the hybrid course as well as the traditional one in the same quarter at the same time of day and students used the same textbook as in the traditional model. Furthermore, the Web-based class used CyberStats, an online introductory statistics course, containing basic text, interactive applications, interactive practice problems and self-assessment tests.

The students had to fulfill assignments each week in both textbook and CyberStats. The hybrid class in the Ward (2004) study met once a week for seventy-five minutes. The time was used to discuss problems and to take tests and quizzes. Two class sessions were scheduled in the computer lab with identical activities as in the traditional course. No new material was presented in class, emphasis was put on the material the students had learnt on their own. The course Web site was linked to data and statistics resources and included guidelines and hints, a daily study calendar, a bulletin board for comments, a chat room and course content modules. These modules included links to daily activities with interactive worksheets, applet demonstrations of statistical concepts, review sheets and practice tests with solutions, links to suggested online readings and Power Point textbook reviews as suggested by the study calendar. Altogether, this course was divided into fifty percent of online classes and fifty percent classroom teaching. In addition to the possibilities for communication between student and teacher in the traditional class, online office hours were scheduled via the Internet chat room. Brown (Brown and Liedholm 2002) taught his hybrid course in face-to-face lectures of two class hours a week with a variety of online materials such as a collection of interactive, collaborative practice materials, a set of Power Point slides as a textbook complement and files of repeatable practice quizzes. Also students could watch a video of Liedholm's lectures in an online format that included synchronous viewing of textual material. Independent from the three class designs indicated above there are obviously some potential benefits of hybrid classes for students, as well as for the educational institution. These are that students have the advantages of an online course and face-to-face guidance and support of an instructor, students must learn to work independent in the online environment (with more constructed ideas) but must also use communication skills in an interactive traditional classroom setting (with spontaneous brainstorming); technology reluctant students learn how to navigate online courses but do not entirely give up live class and at last the time aspect allows students to be more flexible and institutions to offer more classes (Ward 2004).

2.3.3 Internet-Based Instruction

In online classes the instruction takes place through Web based components such as chat rooms, threaded discussion groups, Internet activities, videos, slides of course materials and linked resources. This model allows students to work on their own schedule in different locations (compare Ward 2004). Many studies were carried out to develop the typical profile of online course members but with different results, as summarized in Ward (2004). A study at a community college found out that typical online learners are female, Caucasian, twenty-six to fifty-five years old, work full time as professionals and have higher education and family income than their counterparts in traditional classes (Halsne and Gatta 2002), while studies at some state universities found that their online students are actually traditional students who sign up due to scheduling or logistics problems (Ashkeboussi 2001, Utts et al. 2003). Another survey of the National Education Association found out that the composition of students over and under twenty-five years with either a status of full-time or part-time students is equal (NEA 2000). Due to the very different results it is not possible to adapt the online teaching offering to a specific group of students. Unfortunately, there is only one completely virtual class analyzed in the three studies. This was in the course of Brown and Liedholm (2002) and it was developed by professional Web-course producers, designers, programmers and pedagogical experts under the author's direction. It included all the same materials available in the hybrid class.

2.3.4 Comparison of the Three Models

The aspects of interest of these studies are the students' performance and the evaluation of the most effective instruction mode. In their studies Utts et al. (2003) and Ward (2004) found no significant difference in students' performance as measured by grades concerning their compulsory work but hybrid class students did not complete as much extra work beyond the course requirements as traditional students. As it was the first offering of a hybrid class in Utts et al. (2003) the authors see potential for a useful alternative to the traditional model, although the students in this class appeared to be less happy. Also the feedback of the hybrid class students has provided valuable insight for improvement, e.g. it became clear that these students wanted more discussion time with the instructor. It appears that the best use of class time is to review material that the students have already studied during the previous week and that students are able to learn the material with the online format. Nonetheless, the instructor's role has been reduced to answering questions instead of providing guidance, which is less satisfying for a dedicated educator. In Ward (2004) the students of the hybrid class have a more positive attitude.

This difference in attitude seems to be a result of being more responsible for their performance on required components of the course than traditional students. In their study Brown and Liedholm (2002) concluded that the students in the traditional class achieve significantly better results than the students in the online class when it comes to the most difficult material. Indeed, there is no difference in learning the basic concepts in all three modes. The performance could also partly be due to different students' efforts.[7] Live and hybrid class did not provide considerable differences concerning students' performances. Nonetheless, the results may indicate the advantage and importance of the direct student-teacher interaction of live taught classes.

---

[7] Due to a survey about 51 percent of the students in the virtual class ordinarily spent less than weekly three hours on the course, while in the traditional class in average attendace was over 80 percent.

III. Conclusion

Altogether empirical literature suggests that it is recommendable not to give up instructing face-to-face completely. The result goes along with Merisotis and Phipps (1999, p. 31) who concluded that technology cannot replace the human factor in higher education and it seems that technology is less important than other factors, such as learning tasks, learner characteristics, student motivation and instructors. Thus, in institutions that do not require full distance education, where students are able to attend face-to-face-class occasionally, a hybrid model of the course is recommendable. When the benefits of online learning are combined with the versatility and face-to-face contact of a traditional course, students, instructors and institutions of higher education have the "best of both worlds" (Ward 2004). With regard to Utts et al. (2003) the following recommendations for incorporating Web-based learning materials in econometrics and statistics courses can be made. There should be weekly meetings to keep students on track, motivate them to work on the material (e.g. this can be proofed by quizzes) and guarantee the desired interaction of students with a skilled teacher. Online material could be interactive and more engaging than printed texts but, nonetheless, a textbook for the course should be provided. The authors also think that in following years the virtual part of the class will provide savings in instructor time. Now it is the goal to find the right balance between interactive instruction and the traditional lecture-textbook format, which will differ for different institutions, instructors and student audiences. Apart from the discussion of using Web-based components, there are also other aspects which should be observed to make statistics and econometrics classes more effective as pointed out in Garfield (1995). She suggests that statistics teaching can be more effective if teachers find out what they want to teach their students and how they can do it best. The goal is to provide the students with the skills to apply the contents learned in new situations (see also Hansen et al. 2002). Appropriate assessment should be integrated so that teachers and students can proof (already before the exams) whether the learning goals are being achieved. Statistical instructors should consider the implications of research findings and relate them to their own courses, students and resources and they should also continually assess their own theories of teaching with regard to classroom experience.

Therefore they should also try different teaching and grading modes and other activities and examine the results by tests and student's feedback.[8] In any case they should reorganize their time in class, even if this involves reducing contents. They must lecture less, use active learning and focus on problems and issues. Thus, they successive should (i) introduce a concept, (ii) work through an application, (iii) analyze a case, complete an exercise or discuss a current event, always animating students to participate actively (compare Hansen et al. 2002). To sum up, combining the advantages of the hybrid class instruction with appropriate and comprehensible lecture of theoretical topics in combination with their application to real-world issues, teaching statistics and econometrics will be more effective and students will have less problems in learning.

---

[8] Concerning this, there are certain possible classroom presentation styles (e.g. lecture, classroom discussion, media use), classroom activities (e.g. computer labs, classroom experiments and simulations, cooperative learning, small group work) and assignments involving print or computer-accessed materials on real-world issues. There are also different testing and grading methods, such as multiple-choice, short answer questions, essays, different types of writing assignments (term papers, shorter papers, homework/problem sets), class participation, oral presentations or performance in classroom simulations or experiments (compare Becker and Watts 2001).

Bibliography

1. **Ashkeboussi, R. (2001):** A Comparative Analysis of Learning Experience in a Traditional vs. Virtual Classroom Setting. In: Academic Exchange Quarterly, Vol. 5, No. 4, pp. 133-138.

2. **Becker, W. E.; Greene, W. H. (2001):** Teaching Statistics and Econometrics to Undergraduates. In: Journal of Econometric Perspectives – Vol. 15, No. 4, Fall 2001, pp. 169-182.

3. **Becker, W. E.; Watts, M. (2001):** Teaching Economics at the start of the 21st Century: Still Chalk-and-Talk. In: New Research in Economic Education, AEA Papers and Proceeedings, Vol. 91, No. 2, May 2001, pp. 446-451.

4. **Brown, B. W.; Liedholm, C. E. (2002):** Can Web Courses Replace the Classroom in Principles of Microeconomics? In: Teaching Microeconomic Principles, AEA Papers and Proceeedings, Vol. 92, No. 2, May 2002, pp. 444-448.

5. **Case, K. E. (2002):** Reconsidering Crucial Concepts in Micro Principles. In: Teaching Microeconomic Principles, AEA Papers and Proceeedings, Vol. 92, No. 2, May 2002, pp. 454-458.

6. **Garfield, J. (1995):** How Students Learn Statistics. In: International Statistical Review, Vol. 63, No. 1, 1995, pp. 25-34.

7. **Halsne, A. M.; Gatta, L.A. (2002):** Online versus Traditionally-delivered Instruction: A Descriptive Study of Learner Characteristics in a Community College Setting. In: Journal of Distance Learning Administration [Online], Vol. 5, No. 1. (www.westga.edu/%7Edistance/ojdla/spring51/halsne51.html) (last accessed January, 12th, 2006).

8. **Hamermesh, D. S. (2002):** Microeconomic Principles Teaching Tricks. In: Teaching Microeconomic Principles, AEA Papers and Proceeedings, Vol. 92, No. 2, May 2002, pp. 449-453.

9. **Hansen, L.; Salemi, M. K.; Siegfried, J. S. (2002):** Use It or Lose It: Teaching Literacy in the Economics Principles Course. In: Promoting Economic Literacy in the Introductory Economics Course, AEA Papers and Proceeedings, Vol. 92, No. 2, May 2002, pp. 463-472.

10. **Johnson, H. D. (2005)**: Traditional versus Non-traditional Teaching: Perspectives of Students in Introductory Statistics Classes. In: Journal of Statistics Education, Vol. 13, No. 2, 2005, [Online]. (Print from www.amstat.org/publications/jse/v13n2/johnson.html).

11. **Levine, L. (2002)**: Using Technology to Enhance the Classroom Environment. In: T. H. E. Journal [Online], Vol. 29, No. 6, pp. 16-19. (www.thejournal.com/magazine/vault/A3819.cfm) (last accessed January, 12th, 2006).

12. **MacGregor, C. (2001)**: A Comparison of Student Perceptions in Traditional and Online Classes. In: Academic Exchange Quarterly, Vol. 5, No. 4, pp 143-148.

13. **Merisotis, J. P.; Phipps, R.A. (1999)**: What's the Difference? A Review of Contemporary Research on the Effectiveness of Distance Learning in Higher Education, Washington, D. C.: The Institute for Higher Education Policy, pp. 31

14. **The National Education Association and Abacus Associates (2000)**: A Survey of Traditional and Distance Learning Higher Education Members, Washington, D. C.: The National Education Association.

15. **Paulos, J. A. (1995)**: A Mathematician Reads the Newspaper, New York : Anchor Books.

16. Rensselaer Polytechnic Institute (2001): Press Release, The Pew Grant Program in Course Redesign at the Center for Academic Transformation, Renselaer Polytechnic Institute [Online]. (www.rpi.edu/web/News/press_releases/2001/cat.html) (not available anymore, quoted after Ward, B. (2004): The Best of Both Worlds: A Hybrid Statistics Course. In: Journal of Statistics Education, Vol. 12, No. 3, 2004; nonetheless comparable information is available on Rensselaer Polytechnic Institute (2001): Press Release, Center for Academic Transformation Announces $2 Million in Grants to Ten U.S. Universities to Redesign Courses Using Technology, Renselaer Polytechnic Institute [Online]. (http://news.rpi.edu/update.do) (last accessed January, 12th, 2006).

17. **Salemi, M. K.; Siegfried, J. J.; Sosin, K.; Walstad, W. B.; Watts, M. (2001)**: Research in Economic Eduaction: Five New Initiatives. In: New Research in Economic Education, AEA Papers and Proceeedings, Vol. 91, No. 2, May 2001, S. 440-445.

18. **Sowey, E. R. (1983):** University Teaching of Econometrics: a Personal View. In: Econometrics Review, Vol. 2, May, pp. 255-289.

19. **Utts, J.; Sommer, B.; Acredolo, C.; Maher, M. W.; Matthews, H. R. (2003):** A Study Comparing Traditional and Hybrid Internet-Based Instruction in Introductory Statistic Classes. In: Journal of Statistics Education, Volume 11, No. 3, 2003, [Online]. (Print from www.amstat.org/publications/jse/v11n3/utts.html).

20. **Ward, B. (2004):** The Best of Both Worlds: A Hybrid Statistics Course. In: Journal of Statistics Education, Vol. 12, No. 3, 2004, [Online]. (Print from www.amstat.org/publications/jse/v12n3/ward.html).

21. **Yablon, Y. B.; Katz, Y. J. (2001):** Statistics Through the Medium of the Internet: What Students Think and Achieve. In: Academic Exchange Quarterly, Vol. 5, No. 4, pp. 17-22.